Elisabeth Regina Alves C. Silva
Flavio Marinho da Silva
José Gustavo da Silva Melo

Analysing urban evolution using remote sensing techniques

Elisabeth Regina Alves C. Silva
Flavio Marinho da Silva
José Gustavo da Silva Melo

Analysing urban evolution using remote sensing techniques

Spatio-temporal mapping of environmental degradation in urban areas

ScienciaScripts

Imprint

Any brand names and product names mentioned in this book are subject to trademark, brand or patent protection and are trademarks or registered trademarks of their respective holders. The use of brand names, product names, common names, trade names, product descriptions etc. even without a particular marking in this work is in no way to be construed to mean that such names may be regarded as unrestricted in respect of trademark and brand protection legislation and could thus be used by anyone.

Cover image: www.ingimage.com

This book is a translation from the original published under ISBN 978-613-9-65843-5.

Publisher:
Sciencia Scripts
is a trademark of
Dodo Books Indian Ocean Ltd. and OmniScriptum S.R.L publishing group

120 High Road, East Finchley, London, N2 9ED, United Kingdom
Str. Armeneasca 28/1, office 1, Chisinau MD-2012, Republic of Moldova, Europe
Printed at: see last page
ISBN: 978-620-7-96581-6

SUMMARY

Due to the intense urban growth of cities, green areas and water bodies are increasingly susceptible to environmental degradation. In this sense, there has been growing concern about ways to control and mitigate environmental degradation, which requires prior planning of land use. A survey of current land use, which is necessary for planning purposes, can be obtained using multispectral data provided by remote sensing satellites. The aim of this study was to carry out a **spatio-temporal** analysis of **the Olho D'água Lagoon, the main natural coastal lagoon in the state of** Pernambuco, based on products obtained through remote sensing. Two biophysical variables were used in the evaluation: the normalised difference vegetation index (NDVI) to assess the chlorophyll content of the vegetation and the soil-adjusted vegetation index (SAVI) to identify areas of degradation in the area. Three Landsat satellite images from 1989, 2000 and 2010 were used. The study showed that vegetation with considerable chlorophyll content showed a considerable decrease from 1989 to 2010 due to urban densification in recent decades. The NDVI result was corroborated by the SAVI results, which indicated an increase in degradation in the area surrounding the Olho **D'água** Lagoon.

Keywords: Urban densification; Vegetation indices; Environmental changes; Remote sensing.

SUMMARY

CHAPTER 1

INTRODUCTION

In Brazil, bodies of water such as rivers, lakes and reservoirs are often not considered reasons for attraction and conservation in a city's landscape, despite global concern about environmental issues. This is because sanitation practices are still in their infancy as a form of intervention in the landscape, affecting the issue of sanitation and urban drainage. And the inappropriate way in which water bodies are used in the urban environment is reflected in the quality of life of the population, leaving aside the potential that the natural environment has to contribute to a better, healthier and more liveable city for the population (TENÓRIO, 2013).

In nature, the permanence of water resources, in terms of the flow regime of streams, creeks and rivers, as well as the quality of the water that emanates from the sub-basins, stems from natural control mechanisms developed over the course of evolutionary processes in the landscape, which constitute the so-called services provided by the ecosystem. One of these mechanisms is the close relationship that exists between forest cover and water, where the river-forming springs are located. This natural balance has been constantly altered by man through deforestation, the expansion of agriculture, the opening of roads, urbanisation and various other processes of anthropogenic transformation of the landscape, which alter biogeochemical cycles and the water cycle (XAVIER et. al., 2009).

For this reason, the need to adopt effective water resource management tools to avoid or at least mitigate environmental impacts on water resources requires an understanding of the dynamics of natural processes related to the various factors that make up the different terrestrial landscapes. From this perspective, an understanding of the patterns relating to climate, soil, vegetation, hydrography and land use are important components for drawing up an integrative analysis that aims to understand the natural landscape and man's role in transforming the different environments (SILVA, 2016).

Tundisi (2009) points out that the multiple uses of water, the dumping of waste into rivers, lakes and reservoirs, as well as the destruction of salt water and riparian forests have led to continuous and systematic deterioration, as well as high losses in the quantity and quality of water available to society. With the excessive paving of cities and less and less vegetation on the ground, water flows faster, carrying polluting materials into rivers and seas, eroding soils and hindering the natural recharge of aquifers.

On the Brazilian coast, coastal lagoons are very abundant and range from small depressions filled with rain and/or sea water, of a temporary nature, to large bodies of water, being of great environmental importance and constituting interface regions between coastal zones, inland waters and coastal marine waters (ESTEVES, 1998).

In Pernambuco, the Olho D'água Lagoon is the state's main natural coastal lagoon and one of the largest urban lagoons in Brazil. Located in the municipality of Jaboatão dos Guararapes/PE, in the Metropolitan Region of Recife, with population densities in the neighbourhoods of Piedade, Candeias and Barra de Jangada, all in the same municipality, the lagoon has a water mirror of approximately 3.75 km^2 (FRANÇA et al, 2012). According to Macedo (2010), the lagoon is approximately 2.4 km from the coast, from the point on its eastern shore closest to the coast, and corresponds to an extremely shallow lagoon system of several smaller lagoons, forming the micro-basin of the Jaboatão River and the main artificial canals (the Setúbal Canal and the Olho D'água Canal, which has an outlet to the Jaboatão River). Surrounding the lagoon are areas with sandy and clay soils, where the restinga vegetation has almost all been removed, and today has a vegetation cover made up of various types of trees, shrubs and undergrowth.

The environmental importance of the Olho D'água Basin and, consequently, the interest in protecting the lagoon itself, is primarily due to the fact that it is the only remaining restinga area in the state. In the last ten years, however, there has been strong economic and social pressure, mainly due to the increasing occupation of the area of the basin and even around the lagoon, for the construction of housing, housing developments and buildings. As a result, many of the natural resources in the basin, such as water, soil, native fauna and flora, have been deteriorating, resulting in serious sanitary-environmental, public health and socio-economic problems (TENÓRIO, 2013).

In this sense, remote sensing techniques have become an important tool for environmental monitoring and are increasingly used in environmental analyses, as they comprise procedures capable of generating maps and final products with rich content, associated with various elements and themes. The images obtained through these techniques provide a multi-temporal overview of environments and their transformation, highlighting impacts caused by natural and anthropogenic phenomena or other alterations to the earth's surface (FLORENZANO, 2005).

According to applied studies on remote sensing (WENG; QUATTROCHI, 2006; MELESSE et al., 2007), most of the work carried out in recent decades has focussed on natural environments, while applications of these technologies in urban areas are relatively recent. With the advent of medium- and high-resolution images and more robust techniques, remote sensing has been widely used to map types of land use and cover and to detect the elements that make up the intra-urban space.

In the case of land use and land cover mapping, satellite images make it possible to acquire and exploit certain data and information for analysing environmental quality, which can subsidise territorial planning and management. The level of detail and precision obtained will depend on the characteristics of the orbital sensors, the techniques used to process the images and knowledge of the study area (CARDOSO and AMORIM, 2014).

Quantification, risk assessment and environmental monitoring can be carried out using biophysical parameters (such as vegetation indices) and physical parameters (albedo, temperature, emissivity, among others) obtained from orbital images to determine changes in the surface. Vegetation indices, such as the normalised difference vegetation index (NDVI), can help monitor certain biomes, as well as making it possible to analyse the natural and man-made landscapes of their various ecosystems, thus improving understanding of their structure, functioning and ecological function (LOPES, 2010).

The aim of this study is therefore to carry out a spatio-temporal analysis of the Olho D'água Lagoon based on the NDVI and SAVI vegetation indices, applied to satellite images from the Landsat-5 TM sensor. The aim is to develop products that can be used for various purposes, such as monitoring the growth of the urban area and the dynamics of vegetation in the area.

1.1 Problematisation

In recent decades, the Olho D'água Lagoon has seen a loss of vegetation due to urban densification in the area. For this reason, there is a need to study how this process has affected vegetation dynamics in the area.

1.2 Justification

The Olho D'água Lagoon is the only urban sandbank formation in Pernambuco. It is also considered to be the largest lagoon on the urban perimeter in Brazil and is therefore an Environmental Preservation Area. However, in recent decades there has been an intense process of urbanisation, both vertical and horizontal around the area, which has altered the local landscape and the dynamics of the vegetation. As such, more studies assessing both spatial and temporal changes in the region and monitoring changes to the natural components present in the landscape are of great value.

1.3 Hypothesis

The hypothesis is that the process of urban densification is the biggest factor in the degradation of the Olho D'água Lagoon area.

CHAPTER 2

OBJECTIVES

2.1 General

Carry out a spatio-temporal analysis of the Olho D'água Lagoon based on products obtained through remote sensing.

2.2 Specific

E Estimation of the Normalised Difference Vegetation Index (NDVI);

E Estimation of the Soil Adjusted Vegetation Index (SAVI);

Analysing urban development in the Olho D'água Lagoon.

CHAPTER 3

THEORETICAL FRAMEWORK

3.1 Definition of Coastal Lagoons

Lagoons are defined as shallow bodies of salty, fresh or brackish water, where solar radiation can reach the sediment, allowing aquatic macrophytes to grow throughout their length (ESTEVES, 1998). This is a similar concept to that presented by Guerra & Guerra (2011), which states that lagoons are depressions of small depths, with various shapes, especially circular ones. Esteves (1998) classifies the compartments of this lake ecosystem as a water-air interface region.

It is an ecosystem commonly used by the population to obtain food, for their own survival or for that of their dependents who live further away. This explains why the surrounding areas have been occupied and used to carry out a series of actions and activities that are fundamental to life. However, these uses and occupations have their limits, notably due to the vulnerability of this ecosystem (ASSIS, 2013).

For Loureiro et al, (2006), coastal lagoons are bodies of water connected to the ocean and formed as a result of the rise in sea level during the Holocene/Pleistocene and the construction of sandy restingas through marine processes, partially or totally isolating the lagoon bodies from the ocean. According to Tundisi & Matsumura-Tundisi (2008), a coastal lagoon is defined as a shallow lake or body of water connected to a river or the sea. According to Kjerfve (1984), the definition of a coastal lagoon is a shallow, coastal body of water separated from the ocean by a barrier, connected at least intermittently to the ocean by one or more restricted connections and normally orientated parallel to the coast.

Lagoons account for 15 per cent of the world's coastal zone and are among the most productive ecosystems in the biosphere. These bodies of water serve various anthropogenic activities related to food, energy, transport, recreation and urbanism and their natural balance can be easily disturbed, often irreversibly and always accompanied by socio-economic problems (BARROSO et al, 2000).

According to Nascimento (2010), lake ecosystems such as coastal lagoons are of great importance, both for the various activities they offer to man and for the provision they make for the balance of the ecosystem present there. Understanding the dynamics of these coastal lagoons is important not only for understanding the metabolism of tropical lake environments, but also for supporting programmes aimed at their conservation and rational use. According to Esteves (1998), coastal lagoons contribute to:

o Maintaining the water table;

And local climate stability.

The Coastal Zone is responsible for many "ecological functions", such as: preservation from flooding, saltwater intrusion and coastal erosion; protection from storms; the recycling of nutrients and polluting substances; the provision of habitats and resources for a variety of directly and indirectly exploited species. The most important ecosystems in this zone are estuaries, mangroves and coastal lagoons, as well as marshes and coastal wetlands (NASCIMENTO, 2010).

Planning spaces on the banks of bodies of water is one of the great challenges facing contemporary urban environmental management. Addressing the issue involves confronting the dichotomous relationships involved. On the one hand, the proximity of water has guided the structuring of the city throughout history and spaces on the banks of bodies of water have always performed important urban functions. On the other hand, riparian zones - which are the ecosystems specific to areas on the banks of bodies of water - are the most dynamic areas of the river basin in hydrological, geological, geomorphological and ecological terms (LIMA, 1996), performing essential environmental functions.

The banks of bodies of water are defined as Permanent Preservation Areas by the Brazilian Forestry Code. This is a legal mechanism created to protect environmentally sensitive areas such as steep slopes, hilltops, mangroves, dunes, riverbanks and lakes. The concept of APP includes what is known as the principle of intangibility: the prohibition of any form of use and occupation (MELLO, 2009).

3.2 Urbanisation Concept

One of the major problems facing national and state territory in recent times is disorganised urban sprawl. This activity not only harms the quality of life of all individuals, but also has a direct impact on the environment, leaving the fauna and flora of the most diverse biomes, and their respective areas of coverage, reduced. This accelerated urbanisation process in Brazilian cities has resulted in the occupation of various areas without adequate planning to minimise or avoid the degradation of occupied areas (NOVAES, 2010).

This predatory, disorganised and unsustainable model of urban sprawl has lasted throughout Brazilian history, since the industrial revolution, when cities, unable to cope with the huge migration of the rural exodus and lacking adequate planning, began to expand their areas unconditionally. According to Muller & Martine (1997) quoted by Martins (2012): "The rural exodus, triggered an intense rural-urban population migration, leading to the disorganised arrangement of the population in the hitherto existing cities and/or contributing to the emergence of new population clusters."

According to Mucelin and Bellini (2008), one factor that has contributed greatly to environmental impacts is the constant creation of cities and the growing expansion of urban areas, usually due to migration between poor regions and large metropolises.

According to VILLAÇA (2001), the preferred directions of urban expansion are a subject close to the hearts of geographers, urban planners and property developers. It is common, for example, for masterplans to "predict" the directions "in which the city should grow" and for proposals to be created accordingly. However, it is quite common to identify areas in the process of urban expansion without the slightest proper planning.

According to VILLAÇA (2001), the dynamics of a regime of accumulation determine the way in which built space is produced and transformed. However, this foundation does not evaluate the possible impacts on the environment, given that population growth is taken into account as the predominant factor for urban expansion.

The development that has taken place in Brazil as a result of developmentalist policies has led to the accelerated growth of Brazilian cities, the result of which has been their heterogeneous configuration (MARTINS, 2012). This process ends up not assessing the environmental risks that ultimately result in major impacts on the environment, as described by OLIVEIRA (2009), who says that the result of this and other processes (e.g. industrialisation of cities) is air and water pollution, open sewage disposal, slum settlements, occupation of hillsides and risk areas. The general effect of urbanisation has been the intensification of urban pollution. This has affected resources - water, soil and air - and the agricultural environment (GEHLEN, 2010).

The Olho D'água Lagoon is located in the municipality of Jaboatão dos Guararapes in the state of Pernambuco. As part of the city's structure, it has suffered from the urban agglomerations around it: water pollution, contamination and degradation of its banks, and the de-characterisation of its fauna and flora. The Olho D'água Lagoon is the only urban restinga formation in Pernambuco. It is also considered to be the largest urban lagoon in Brazil, almost three times the size of the Rodrigo de Freitas Lagoon in Rio de Janeiro. Although it is part of this scenario of devaluation and environmental imbalance due to anthropic activities, the Olho D'água Lagoon is part of the estuarine ecosystem of the Jaboatão River and is fundamental to the urban hydrological cycle, and is the main element of its micro-basin, which bears the same name (TENÓRIO, 2013).

However, there is a problem of managing the landscape of this ecosystem, which first needs to be understood within the urban processes of which it is a part. Its relationship with the urban environment is not linked to the formation of the city, but is directly linked to the basic functions that a lagoon performs within the urban ecology: flood dampening, receiving rainwater from a river basin, climate mitigation and subsistence economic activities (TENÓRIO, 2013).

According to Freitas (2008), the urban-climate system is an open and dynamic system in which its elements (temperature, humidity, rainfall and ventilation) are constantly modified, mainly by anthropogenic factors, which are essentially local, such as urban morphology, wind corridors, vehicle traffic, industrial activities, landfills, among others. This contributes to the formation of microclimates in the urban environment. Also according to the author, large bodies of water influence climatic elements that often act as wind redirectors, also contributing to microclimates. In this sense, vegetation cover plays an important role, both in surface runoff - part of the rain that runs off the surface of the ground - and in base runoff - the result of water percolating from the ground - where it moves at low speeds, feeding rivers and lakes. The removal of vegetation cover reduces the observed time interval between rainfall and the effects on watercourses, decreases the water retention capacity of river basins and increases the flood peak. In addition, vegetation cover limits the possibility of soil erosion and minimises sediment pollution of watercourses (BENZERRIL Jr., 1993).

3.3 Characterisation of the Olho D'água Lagoon

The municipality of Jaboatão dos Guararapes, an important territory in the Recife Metropolitan Region, is crossed by the Jaboatão River, which, as it runs through the city, is polluted and contaminated at every turn and with every agglomeration that settles in its waters. Its waters are currently unfit for human consumption and supply. Also known for its coastline and estuarine system, Jaboatão dos Guararapes faces many conflicts with its urban waters. Urban waters are linked to landscape conservation problems in the Recife Metropolitan Region. Drainage and sanitation problems are challenges to the constitution of a water landscape that remains in people's imaginations related to well-being and pleasantness in Greater Recife (TENÓRIO, 2013).

According to Assis et al (1997), from the 1940s onwards the municipality of Jaboatão dos Guararapes underwent an accelerated process of demographic densification. Between 1980 and 1990, the municipality's population doubled, with a growth rate of 478% in 30 years. It was during this period that the population began to concentrate predominantly on the seafront (ASSIS ET AL, 1997). The sharp increase in buildings on the seafront, which initially catered for a population with medium to high purchasing power, attracted construction labour, which ended up living in the adjacent areas, precisely around the Olho D'água Lagoon, adding to the population that already existed there (EMDEJA, 2003).

The Olho D'água Lagoon, located in Jaboatão dos Guararapes, is the main natural lagoon on the coast of Pernambuco, and one of the largest in an urban area in the Northeast and in the country. Its water mirror covers an area of 3.75 kilometres^2 and is approximately 3.5 kilometres long and 1.9 kilometres wide. The lagoon itself is also unique in that it is a water link between two estuaries in the Recife Metropolitan Region. Through the Olho D'água Canal, it connects to the estuary formed by the mouths of the Jaboatão and Pirapama rivers in Barra de Jangada, and through the Setúbal Canal, to the estuary of the Pina river, at the confluence of the Tejipió, Jordão and Capibaribe rivers in Recife (SANTOS and KATO, 1997).

The Olho D'água Basin is the only remaining restinga area in the state of Pernambuco. In addition, the area around the lagoon is part of an important estuarine ecosystem of around 3.7 square kilometres and is therefore an area of environmental preservation. However, the lack of planning has led to a disorderly occupation of the land, with many land problems, as well as a lack of urban infrastructure. Problems of rubbish, water and beach pollution, domestic and industrial sewage, flooding, invasions and illegal construction are all visible (TENÓRIO, 2013).

Coastal lagoons are sensitive ecosystems with high productivity. The Olho D'água Lagoon, being a coastal lagoon, is classified as eutrophic in terms of productivity (LEAL, 2002). According to Tenório, due to the occupation of the surrounding areas, lack of infrastructure, silting up, degradation, among other processes that de-characterise its ecosystem, the biodiversity of the lagoon and its surroundings is decreasing day by day and the process of eutrophication of the ecosystem is accelerating, compromising its environmental quality.

Despite its privileged location and great economic and tourist potential, the Olho D'água Lagoon suffers from degradation and abandonment by the government. Despite its unique and exuberant beauty, the lagoon is currently suffering from sewage pollution, irregular occupation, fires, the killing of native animals and silting up. Recent studies show that the level of pollution is so great that the lagoon is even losing its natural depth. The lagoon used to be around 8 metres deep. Today it is less than 80 centimetres deep in some places (FERNANDES, 2011).

Siltation, which has brought sediment into the lagoon, has affected all the biodiversity in the area. There are almost no fish and the few species that remain are threatened. The natural vegetation suffers from intermittent fires and the mangrove area is almost extinct. Native animals are disappearing and birds no longer frequent the lagoon as they once did. The consequences of all this destruction are mainly felt during the winter, when all the neighbourhoods in the southern part of Jaboatão, from the richest to the humblest, are completely flooded. With the lagoon silted up and nowhere to drain, the rainwater invades the urban area, bringing disease, damage and inconvenience to the population of the entire south of Jaboatão (FERNANDES, 2011).

According to TENÓRIO (2013), there is a need for integrated measures to preserve, recover and improve the environmental conditions of the Olho D'água Basin. As part of the Olho D'água Lagoon Revitalisation Programme, the Jaboatão dos Guararapes City Council (PMJG) has established priority goals that include sewage disposal and treatment, macro-drainage, urban cleaning, urban structuring and development, landscape management and treatment, flora and fauna preservation, and environmental education.

The lagoon ecosystem establishes exchanges with the urban system that go beyond its physical limits. It participates in the urban climate system, establishing a more local relationship within its micro-basin, in the municipality of Jaboatão dos Guararapes, but there is also a relationship between this lagoon and the Metropolitan Region of Recife, based on the system of public open spaces and the interests in the area situated in a strategic location (TENÓRIO, 2013).

In the case of the Olho D'água Lagoon, this is intensified by its coastal location, which is therefore influenced by the sea. In addition to the bodies of water, the diversity of vegetation along its banks is remarkable, despite the replacement and landfill generated by human occupation. On a local scale, vegetation has a major influence on environmental elements and may be the main factor responsible for the formation of a microclimate (FREITAS, 2008:78). According to the Jaboatão dos Guararapes City Hall (1996:4), at the end of the 1960s the lagoon's environment was balanced, and there were still no urban pressures or construction densification processes on its banks, as there would be at the end of the 1970s. Its surroundings had vegetation made up of "a juncai association and a hydrophilic meadow of Cyperaceae and Grasses".

In turn, coastal lagoons and lakes represent the ideal environment for some species, defend the land from the aggression of salinity, preserve coastlines from marine erosion, act as climate regulators, both in the coastal zone and inland; as well as being of great economic, social and tourist importance for various municipalities, when used rationally. However, these water bodies are little known ecologically (CAVACA, 1992; OLIVEIRA et al., 2000).

3.4 Contextualising Remote Sensing and the Environment

Remote Sensing is defined in different ways by various authors, the most usual definition being that adopted by Avery and Berlin (1992) and Meneses (2001): "A technique for obtaining information about objects through data collected by instruments that are not in physical contact with the objects under investigation". For Jensen (2007), Remote Sensing (RS) can be defined as the art and science of obtaining information about objects without direct physical contact with the object.

Remote Sensing can also be defined as the joint use of modern sensors, data transmission equipment, data processing equipment, aircraft and spacecraft with the aim of studying the interactions in the earth's environment, without direct physical contact with the features, between electromagnetic radiation and the substances that make up planet Earth in its various manifestations (NOVO, 1995).

The development of remote sensing techniques has made it possible to acquire a variety of information about the earth's surface, supporting temporal, edaphic and phenological analyses of vegetation (VINAGÓ et al, 2011). Pereira et al (1989) state that a survey of current land use, which is necessary for planning purposes, can be obtained using multispectral data provided by remote sensing satellites, combined with interpretation techniques.

In order to understand changes in a given environment, it is necessary to use satellite images from different dates, thus obtaining a multi-temporal analysis. Studies on multitemporal analysis using satellite images are becoming increasingly popular, serving to monitor urban growth as well as to analyse the evolution of deforestation and agricultural extension, among others (CARVALHO JÚNIOR et al., 2005).

In the case of land use and land cover mapping, satellite images make it possible to acquire and exploit certain data and information for analysing environmental quality, which can subsidise territorial planning and management. The level of detail and precision obtained will depend on the characteristics of the orbital sensors, the techniques used to process the images and knowledge of the study area (CARDOSO and AMORIM, 2014).

Often the processing of images and their conversion into thematic maps through manual interpretation or multispectral classification is a time-consuming job with low precision. This is because the advance of remote sensing techniques has provided a large amount of information obtained by orbital sensors, and when it comes to urban areas the diversity of targets implies spectral confusions that reduce the consistency of the cartographic product with reality (XU, 2007; ZHA et al., 2003).

On the other hand, in order to automate the process of mapping the earth's surface based on the unique spectral response of the different targets to a given region of the electromagnetic spectrum, appropriate combinations between the bands of satellite images have served as the basis for the development of a set of radiometric indices that seek to discriminate between the main types of land use and cover (ZHA et al., 2003; PINHEIRO; LARANJEIRA, 2013).

Radiometric indices are measures capable of identifying and highlighting certain types of information in satellite images, such as built-up areas, vegetation cover, watercourses, exposed soil, among others, and normalisation helps to reduce noise and lighting effects (FRANÇA et al., 2012).

In the case of land use and land cover mapping, satellite images make it possible to acquire and exploit certain data and information for analysing environmental quality, which can subsidise territorial planning and management. The level of detail and precision obtained will depend on the characteristics of the orbital sensors, the techniques used to process the images and knowledge of the study area (CARDOSO & AMORIM, 2014).

According to Ponzoni (2001), the appearance of vegetation cover in a given remote sensing product is the result of a complex process involving many environmental parameters and factors. What is effectively measured by a remotely-sited sensor from a given vegetation (target) cannot be explained solely by the intrinsic characteristics of that vegetation, but also includes the interference of various other parameters and factors such as: the radiation source, atmospheric scattering, the characteristics of both the leaves and the canopy, soil moisture levels, interference from soil reflectance, shade, among others.

The use of vegetation indices such as the Normalised Difference Vegetation Index (NDVI), Soil-Adjusted Vegetation Index (SAVI) and Leaf Area Index (LAI) makes it easier to obtain and model biophysical plant parameters such as leaf area, biomass and percentage of ground cover, especially in the infrared region of the electromagnetic spectrum, which can provide important information on plant evapotranspiration (JENSEN, 2009; EPIPHANIO et al., 1996).

Vegetation index modelling is based on the opposite behaviour of vegetation reflectance in the visible region, i.e. the higher the plant density, the lower the reflectance due to absorption of radiation by photosynthetic pigments and the higher the plant density, the higher the reflectance due to scattering in the different layers of the leaves (BORATTO E GOMIDE, 2013).

However, the indices have been identified as indicators of vegetation growth and vigour and can be used to diagnose various biophysical parameters with which they have high correlations, including leaf area index, biomass, percentage of ground cover, photosynthetic activity and productivity (PONZONI, 2001). These indices have been successfully used to monitor changes in vegetation on a continental, regional and global scale (BANNARI et al, 1995 cited by BARBOSA, 2006). This success is due to the differential reflectance of chlorophyll in the visible and infrared wavelengths (BARBOSA, 2006).

CHAPTER 4

MATERIALS AND METHODS

4.1 Characterisation of the Area

The Olho D'água Lagoon, also known as the Náutico Lagoon, is located in the municipality of Jaboatão dos Guararapes - PE, in the Metropolitan Region of Recife.

With a water mirror of approximately 400 hectares, it is the largest sandbank lagoon in an urban area in Brazil, two and a half times larger than the Rodrigo de Freitas Lagoon in Rio de Janeiro (FERNANDES, 2011).

The Olho D'água Lagoon is unique in that it is a water link between two estuaries in the Recife Metropolitan Region. Through the Olho D'água Canal, it connects to the estuary formed by the mouths of the Jaboatão and Pirapama rivers, in Barra de Jangada, and through the Setúbal Canal, to the estuary of the Pina river, at the confluence of the Tejipió, Jordão and Capibaribe rivers, in Recife.

The lagoon's catchment area is 33.5 km^2 and covers the municipality's densely populated and increasingly densely populated area, from the seafront (Piedade, Candeias and Barra de Jangada beaches) to the BR-101 South motorway, in an east-west direction, and from the border with the municipality of Recife to the estuarine zone of the Jaboatão River, in a north-south direction. The estimated population of the basin is around 200,000 inhabitants (SANTOS and KATO, 1997) (Figure 1).

Figure 1 - Location of the study area.
Source: Adapted from Google Earth, 2018.

The lagoon is located in the region known as the "Axis of Development", between the cities of Recife and the Port of Suape (Ipojuca - PE), and therefore has great tourist (as yet unexploited), economic and property potential. According to a survey carried out by the State Secretariat for Cities, 35,000 people live on the shores of the lagoon and another 150,000 within a radius of 2 kilometres (FERNANDES, 2011).

Geologically, the Jaboatão River Basin is mostly made up (90%) of Precambrian crystalline rocks and, secondarily, of Tertiary (barrier formation) and Quaternary sediments (alluvium and beach sediments). The basin's geomorphology consists of 2 main units: The Pliocene surface and the coastal plain. The first unit corresponds to a pediplano sloping from west to east, at elevations of 200 to 400 metres, and is highly dissected by the erosion cycle, which is predominantly linear in action, imposed by the existing hydrographic network. The second, bordering the coast, has elevations of less than 100 metres and its formation is associated with variations in sea level since the end of the Tertiary (GOMES et al., 2003).

According to GOMES et al., 2003, marine transgressions and regressions have caused erosion during periods of high seas and siltation during periods of low seas, clogging up sediments of a fluvial marine nature and coastal strands such as those surrounding the Olho D'água Lagoon in Jaboatão dos Guararapes. The hydrophilous and shrub vegetation is mainly located in the Olho D'água Lagoon area (Figure 2).

According to Braga and Ricardo (2012), the geomorphological fate of lagoons is siltation. It is independent of human influence. It's a process. It's endorheic, you have drainage that flows into it and all the drainage drags sediment, and the sediment doesn't leave, it just stays (so the tendency is to silt up). The eutrophication of a lagoon is a natural process. However, siltation in the Olho D'água Lagoon is aggravated by the occupation processes around it, the alterations to its ecosystem and the constant dumping of organic matter into its waters. This ecosystem is fragile and its imbalance causes damage not only to its biological functioning, but also to the urban environment in which it is located, such as flooding and changes in the urban microclimate (BRAGA, 2012).

Sand is extracted from riverbeds, the Olho D'água Lagoon and river and marine terraces. Clay is removed from the barriers along the BR-101 motorway. Mechanised or semi-mechanised extraction is carried out by medium to large companies, generally with legal authorisation to carry out the activity. The most common substances exploited by these companies are kaolin, sand, clay, granite/gneiss for the production of gravel and mineral water (GOMES et al., 2003).

The environmental impacts caused by these mining operations are far greater than those caused by informal activities. Among the risks related to mineral extraction are: collapse of blocks of rock; risk of unconsolidated material slipping, which can affect nearby residents or even the people directly involved in the extraction (more common in informal activities where no measures are taken with regard to safety at work); silting up of rivers, lagoons and river channels; atmospheric pollution due to the emission of particles into the air; soil erosion and the formation of gullies; contamination of the water table and surface water; the disappearance of native species (flora and fauna) when the vegetation is removed and visual pollution caused by the degradation of the hills. Losses occur in quality of life, health or even the preservation of human life, surface and underground water quality, natural landscapes and in places where urban centres could be established (GOMES et al., 2003).

Leal (2002) recorded the vegetation formations found in the early 2000s (Figure 2), i.e. different from those identified in studies from the 1960s, due to the process of anthropisation around this ecosystem. Remnant formations typical of restinga, mangroves and hygrophilous associations were found, as well as areas with crops or anthropised. The author also noticed an intense anthropogenic modification of the vegetation compared to the formations identified in the 1960s.

Since the 2010s, changes in vegetation formations have been observed as a result of anthropogenic actions, whether it be replacement for planting crops or suppression followed by landfill to make way for occupation. The banks of the lagoon have been urbanised (Figure 2), and as a result there is direct dumping of rubbish and sewage, which further weakens this ecosystem, making it more vulnerable to risks, both in terms of its ecological functioning and for the local population, which suffers from flooding (TENÓRIO, 2013).

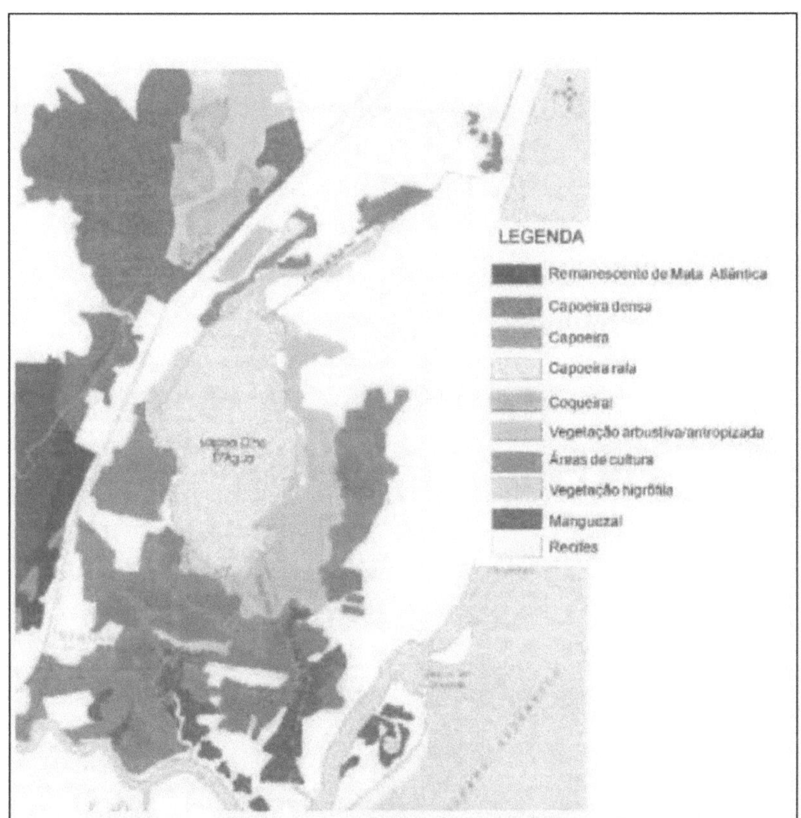

Figure 2 - Map of the vegetation around the lagoon in 2002.
Source: Leal (2002).

According to Tenório (2013), the east side of the Olho D'água Lagoon is home to the main roads along the coastline, which connect the municipality to the Boa Viagem neighbourhood in Recife and the municipality of Cabo de Santo Agostinho (Figure 3). The connection to Recife is made by the avenues Bernardo Vieira de Melo and Airton Senna da Silva, parallel to the beach line with the municipality of Cabo de Santo Agostinho, there is a connection to the south by the Paiva Bridge, linking Jaboatão dos Guararapes to Cabo de Santo Agostinho by the coast (Figure 3).

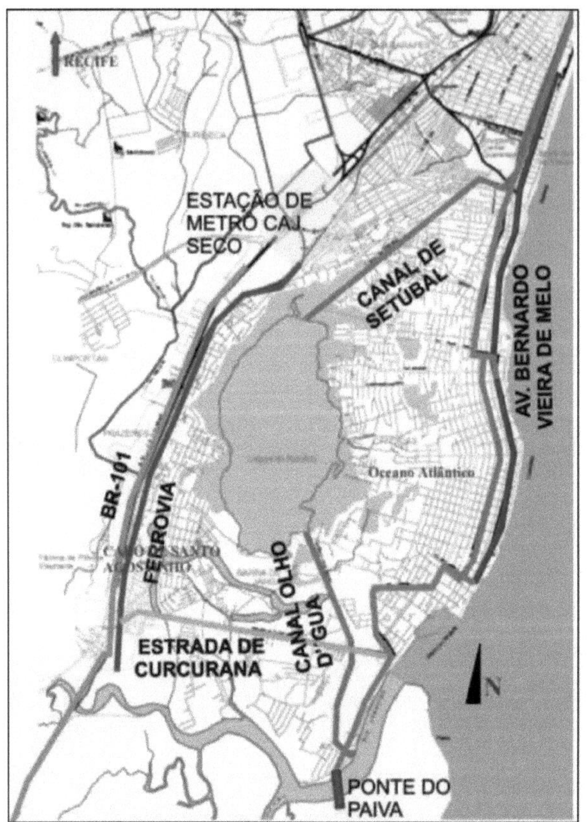

Figure 3 - Road map of the municipalities surrounding the Lagoon.
Source: Tenório, 2013.

The Olho D'água Lagoon covers an area of 375 hectares, is a drainage area and acts as a balancing point for flooding in the region, as it is located on a slope. The city of Jaboatão dos Guararapes has entirely absorbed the Olho D'água Lagoon within its territory, and with the city's urban expansion towards the coast and its conurbation with Recife, many communities have settled without any regulations or infrastructure on the banks of the Lagoon. This disorderly occupation of an area known as baixio (lowland), a space for water and with points where the water table rises, has greatly damaged the ecosystem of the Olho D'água Lagoon, which has caused inconvenience for these communities. So within the systems dealt with here is also society as a system, which interacts with the lagoon maintaining relationships of survival, leisure, contemplation, 'transit', among others (TENÓRIO, 2013).Due to the spatial dimension of the lagoon, its surroundings present distinct social and economic dynamics. The differences can be identified by the pattern of occupation of the area and the types of buildings, as well as the urban infrastructure and road system (Figure 4).

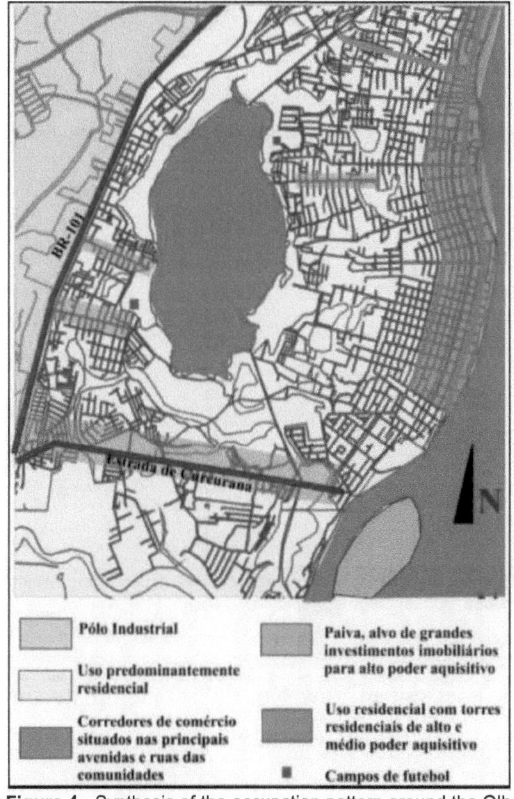

Figure 4 - Synthesis of the occupation pattern around the Olho **D'água** Lagoon.
Source: Tenório (2013).

The Lagoon ecosystem is in the Water Bodies Conservation Zone (ZCA) and the Olho D'água Canal, which connects it to the estuary, is in the Estuarine Protection Zone (Figure 5).

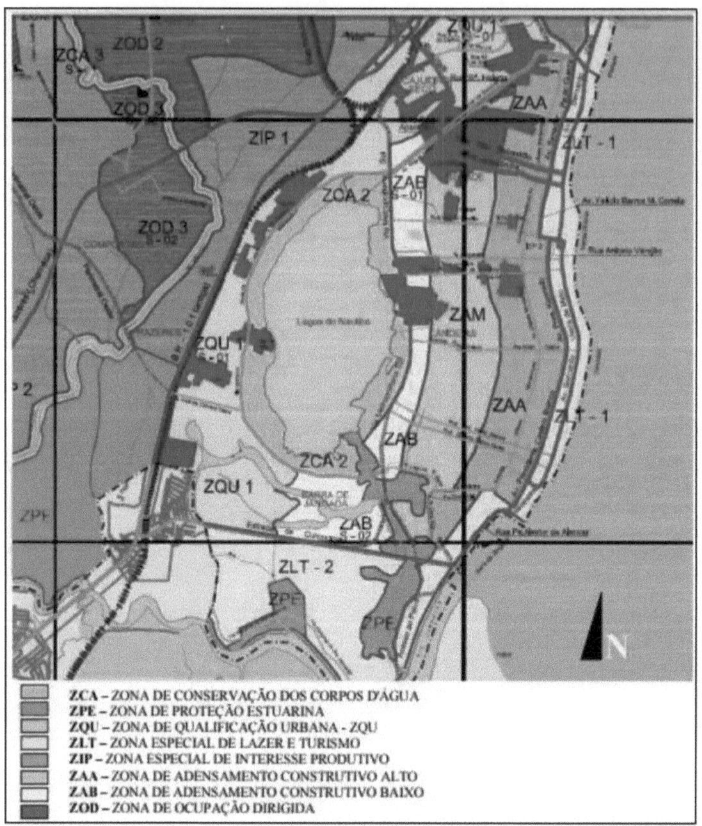

Figure 5: Macrozoning section of the Master Plan for the municipality of Jaboatão dos Guararapes, which places the Olho **D'água** Lagoon in the **Water Bodies** Conservation Zone.

Source: Master Plan of the municipality of Jaboatão dos Guararapes (2006).

As you can see, the guidelines set out in Law No. 068 contemplate the conservation of the Olho D'água Lagoon ecosystem and the reversal of the current fragilities. However, so far what has been done under these guidelines is to remove the population bordering the east bank (TENÓRIO, 2013).

4.2 Satellite Image Processing

4.2.1 Space-Time Analysis

Initially, the study area was delimited. Images from 10-07-1989, 09-08-2000 and 06-09-2010 from the Landsat-5 satellite were interpreted and analysed, downloaded free of charge from the website of the National Institute for Space

Research (INPE) and processed at the UFPE Remote Sensing and Geoprocessing Laboratory (SERGEO), which holds the licence for the software used in this work.

4.2.2 Image processing and *layout* assembly

Initially, all the images were recorded from points collected in the field. The images were downloaded from the INPE website. To process the Landsat-5 satellite images, models were created using the Model Maker tool in the ERDAS Imagine 9.3 software licensed from the Department of Geographical Sciences at the Federal University of Pernambuco.

4.2.3 Radiometric calibration

The set of radiance or radiometric calibration is obtained using the equation proposed by Markham and Baker (1987) (Equation 1):

$$L\,\lambda i = \alpha t + \frac{bt - \alpha t}{255}\,ND \quad (1)$$

Where a and b are the minimum and maximum spectral radiances (1 1 2μm srWm), ND is the pixel intensity (integer between 0 and 255) and i corresponds to the bands (1, 2, ... and 7) of the Landsat 5 and 7 satellite. The calibration coefficients used for TM images are those proposed by Chander and Markham (2003) and Oliveira et al, 2010.

4.2.4 Reflectance

solar radiation reflected by the surface and the flux of incident global solar radiation, obtained using the equation (ALLEN et al., 2002 *apud* OLIVEIRA et al, 2010), (Equation 2):

$$\rho\lambda i = \frac{\pi\,.L\lambda i}{K\lambda i\,.cos\,Z\,.dr} \quad (2)$$

Where λiL is the spectral radiance of each band, λik is the spectral solar irradiance of each band at the top of the atmosphere 12μm, Z is the solar zenith angle and rd is the square of the ratio between the mean Earth-Sun distance (ro) and the Earth-Sun distance (r) on a given day of the year (DSA), (OLIVEIRA et al, 2010; SILVA et al, 2011).

4.2.5 *Normalised Difference Vegetation Index (NDVI)*

The Normalised Difference Vegetation Index (NDVI) is the ratio of the difference between the reflectivities of the near infrared and red bands and the sum of these reflectivities (ROUSE et al., 1973). NDVI is a sensitive indicator of the quantity and condition of vegetation, whose values range from -1 to 1. On surfaces containing water or clouds, this variation is always less than 0.

The Normalised Difference Vegetation Index (NDVI) was obtained from the ratio between the difference in the reflectivities of the near infrared (ρIV) and red (ρV) and the sum between them (TUCKER, 1979 apud TASUMI, 2003): (Equation 3):

(IVP – V) / (IVP +V)   (3)

4.2.6 *Soil Adjusted Vegetation Index (SAVI)*

The Soil Adjusted Vegetation Index (SAVI) is an index that takes into account the effects of exposed soil on the images analysed, to adjust the NDVI when the surface is not completely covered by vegetation. SAVI (Equation 3) was developed by HUETE (1988) as a transformation technique to minimise the influence of soil reflectance on spectral vegetation indices involving the red and near-infrared wavelengths and to more accurately model near-infrared radiance in more open canopies (SILVA et al, 2011).

Although the NDVI index has been widely used, it has some limitations, which affect the results achieved, such as interference due to soil colour and moisture effects. An index was therefore developed that could improve NDVI values without the need for field measurements for each area analysed (Jensen, 2009). To this end, an improved index was developed using a constant, L, as an adjustment factor for the canopy substrate. Thus, the Soil Adjusted Vegetation

Index (SAVI) is expressed as (Equation 4):

$$SAVI = \frac{(1+L)(\rho IV - \rho V)}{(L + \rho IV + \rho V} \quad (4)$$

Where: L is a constant known as the SAVI index adjustment factor, and can take values from 0.25 to 1 depending on the soil cover. According to Huete (1988) a value for L of 0.25 is indicated for dense vegetation and 0.5 for vegetation with intermediate density, when the value of L is 1 for vegetation with low density. If the SAVI value is equal to 0, its values become equal to the NDVI values. Therefore, the most commonly used L value is 0.5.

The constant L can have values from 0 to 1, varying according to the biomass itself. According to Huete (1988) apud Ponzoni and Shimabukuro (2009), the optimum values for L are: L = 1 (for low vegetation densities) L = 0.5 (for medium vegetation densities) L = 0.25 (for high vegetation densities). According to the aforementioned authors, in general the L = 0.5 factor is more commonly used, since it encompasses a greater variation in vegetation conditions. Even so, SAVI is limited by the different biomes and agricultural situations, since the constant values are generalised and do not take into account the specificities of the environments analysed, but only the vegetation density (PONZONI, SHIMABUKURO, 2009).

CHAPTER 5

RESULTS AND DISCUSSION

In this study, the images from 10-07-1989, 09-08-2000 and 06-092010, in sequence as shown below (Figure 6), were georeferenced and processed in ERDAS Imagine 9.3 software to obtain the NDVI for these years, and the layout was prepared in ArcGis 10.2 software. The NDVI estimates the chlorophyll content (greenish pigment) present in the vegetation, so where there is healthy vegetation the NDVI will identify it (Figure 6).

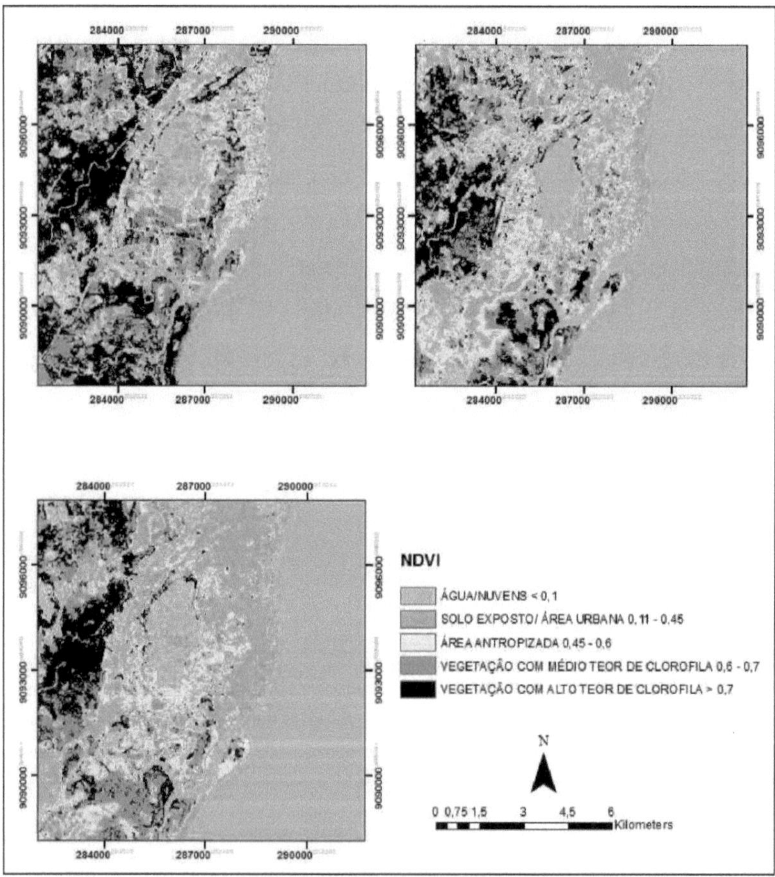

Figure 6. Chlorophyll content in the vegetation of the Olho **D'água** Lagoon.

Source: Prepared by the author (2018).

Thus, NDVI less than 0.1 corresponded to the presence of clouds and water in Figure 4. Over the years there has been a decrease in green areas around the Olho D'água Lagoon, while urban areas have increased in density, occupying spaces that were previously vegetated areas, which can be seen by the increase in the urban sprawl and decrease in the chlorophyll content of the vegetation (NDVI < 0.45). The anthropised areas in the previous figure (0.450.6) were those which, despite being pressured by the surrounding urban densification, still have a certain area of vegetation (Figure 6).

The vegetation with considerable chlorophyll content over two decades showed a considerable decrease from 1989 to 2010. The areas with a predominance of values ranging from 0.6 to 0.7 are precisely those that move further inland from the coast. The same pattern can be found in the vegetation with a high chlorophyll content (NDVI > 0.7), which in the 80s was somewhat predominant in the area. It should be emphasised that the area with the highest chlorophyll content is precisely the one that belongs, according to the Master Plan of the municipality of Jaboatão dos Guararapes, to the Estuarine Protection Zone (Figure 6).

The closer NDVI values are to 1, the denser the vegetation; a value of 0 (zero) indicates an unvegetated surface (ROSENDO, 2005). The highest NDVI values correspond to the highest Digital Numbers (ND), which are related to areas of vegetation with greater vigour. The lowest values correspond to low ND, representing areas of stressed vegetation, much less dense or even bare areas. Jensen (2009) presents some positive and negative points when using NDVI. For this author, the importance of this index centres on two aspects: 1) monitoring seasonal and inter-annual changes in vegetation activity and development and; 2) reducing noise, such as cloud shadows, topographical variations and differences in solar illumination.

Compared to the ecological record made in the 1960s by Coelho, P. A. (1965/6), Leal (1995) observed a marked reduction in its ecological components around the Olho D'água Lagoon, identifying intense urban occupation today. According to the Preliminary Environmental Report of the Environmental Impact

Studies of the Olho D'água Lagoon by the Empresa Municipal de Desenvolvimento de Jaboatão dos Guararapes - EMDEJA (2003), the fauna present in and around the lagoon is characterised by species adjusted to urban conditions or environments close to occupation, i.e. its landscape is both natural and urban.

An index that takes into account soil adjustment (SAVI) was also used to analyse the area over time. This index has the potential to highlight the patches of exposed soil present in the area and better highlight the degradation present in some areas, since it is one of the constituent components of the leaf area index (LAI), which is therefore influenced by the biomass present in the area. The SAVI result is shown below (Figure 7).

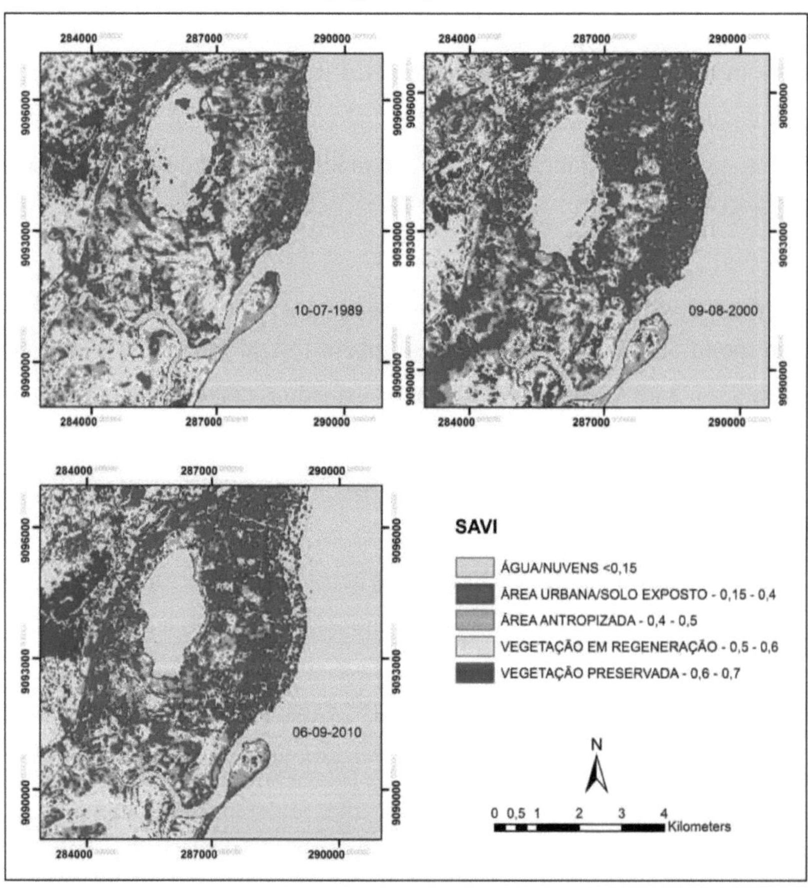

As can be seen in the figure above, the images from 2000 and 2010 showed greater urban densification compared to 1989. The anthropised areas also showed growth compared to the late 80s. Preserved vegetation (SAVI > 0.6) only grew in the area corresponding to the Estuarine Protection Zone (Figure 7).

The soil-adjusted vegetation index (SAVI), which seeks to minimise the effects of soil reflectance by incorporating a factor

adjustment dependent on vegetation density (GILABERT et al, 2002; EASTMAN, 2006).

Below is a Google Earth image of the Olho D'água Lagoon area in 2014 (Figure 8).

Figure 8. Urbanisation pattern of the Olho D'água Lagoon, 2018.
Source: Adapted from Google Earth (2018).

The figure above shows the pattern of urbanisation around the Olho D'água Lagoon and how it has been the main stressor in the area. The green areas still present in the area correspond to environmental protection areas such as the Estuarine Protection Zone provided for in the Master Plan of the municipality of Jaboatão dos Guararapes.

One of the main interventions in the area is currently linked to the Setúbal

Canal, which is 2.9 kilometres long. Jaboatão's Executive Secretary for Maintenance, Manoel Chaves, highlighted the participation of the Municipal Administration in the implementation of the Development Plan, where all the treatment of the silting of this and other canals that end up flowing into it along its course is taking place. "In addition, the Municipal Government is responsible for all the lighting and tree-planting along the road," said the secretary.

Through the municipality of Jaboatão dos Guararapes in Greater Recife, the Olho D'água Lagoon has received R$31 million in funding for works. The money was released by the Ministry of Cities through the Sanitation for All Programme and will be used for paving, drainage and the construction of a 2.8-kilometre cycle path.

According to the Pernambuco Department of Cities, R$14.2 million will be invested in resizing 1.7 kilometres of the Olho D'água Canal. The service will be carried out by macro-draining and widening the channel, which is essential to prevent flooding around the lagoon, also known as Lagoa do Náutico. A further R$18 million will be invested in paving and draining the marginal roads of the Setúbal Canal, from Av. Barreto de Menezes to the vicinity of the lagoon. The project also includes a bidirectional cycle path on one of the banks. Between 2011 and December 2012, Cehab completed the first stage of the lagoon revitalisation project with the delivery of 1,376 housing units, benefiting more than 4,000 families in the area.

CHAPTER 6

FINAL CONSIDERATIONS

The NDVI analysis showed that the vegetation with considerable chlorophyll content over two decades showed a considerable decrease from 1989 to 2010 due to urban densification in recent decades. The area with the highest chlorophyll content belongs to the Estuarine Protection Zone.

The NDVI result was corroborated by the SAVI results. The images from 2000 and 2010 showed greater urban densification compared to 1989. The anthropised areas also showed an increase compared to the late 80s.

From the results analysed, it can be concluded that the intense urbanisation that has taken place over the last few decades is the preponderant factor in the decline of vegetation over the years and that this urban densification in the area tends to put increasing pressure on the vegetation in the Olho D'água Lagoon area, increasing the anthropised areas around it. In this way, the products obtained through remote sensing were useful for a spatio-temporal understanding of the territorial dynamics of the Olho D'água Lagoon.

CHAPTER 7

REFERENCES

ALLEN, R.; TASUMI, M.; TREZZA, R. SEBAL (Surface Energy Balance Algorithms for Land). **Advanced Training and User's Manual** - Idaho Implementation, version 1.0. 2002.

ASSIS, H. M. B. de. **Project to diagnose the physical environment of the Olho D'água Lagoon basin.** Recife: CPRM/PMJG, 1997.

ASSIS, D. R. S.; PIMENTEL, R. M. M.; CASTILHO, C. J. M. Impacts of Urbanisation and Vulnerability of Coastal Lagoons. **Revista Brasileira de Geografia Física** V. 06 N. 02 (2013) 223-232.

BARBOSA, K. M. do N. **Spatial monitoring of biomass and organic carbon of floodplain herbaceous vegetation in Central Amazonia.** Curitiba: Federal University of Paraná, Doctoral Thesis, 131 p., 2006.

BARROSO, V. L., MEDINA, R. S., MOREIRA-TURQ, P.F., BERNARDES, C.M., **Aspectos Ambientais e Atividades de Pesca em Lagoas Costeiras Fluminenses.** Environment in Debates Series, 31. Brazilian Institute for the Environment and Renewable Natural Resources, Strategic Management Directorate - Brasília. Ed. IBAMA, 50p, 2000.

BORATTO, I. M. de P.; GOMIDE, R. L. **Application of the NDVI, SAVI and IAF vegetation indices to characterise vegetation cover in the northern region of Minas Gerais.** Anais XVI Simpósio Brasileiro de Sensoriamento Remoto - SBSR, Foz do Iguaçu, PR, Brazil, 13 to 18 April 2013, INPE.

BENZERRIL JR., P. **Direito de Águas e Meio Ambiente.** São Paulo: Ícone, 1993.

CARDOSO, R. dos S.; AMORIM, M. C. de C. T. Evaluation of the NDVI, NDBI and NDWI indices as tools for mapping land use and cover. **VII National Congress of Geographers.** Vitória-ES. Proceedings. 2014.

CARVALHO JÚNIOR, O.A; GUIMARÃES, R.F; CARVALHO A.P.F; GOMES, R.A.T; MELO, A.F; SILVA, P.A. **Processing and analysis of multitemporal images for the Gorutuba/MG irrigation perimeter.** In: Actas XII Simpósio Brasileiro de Sensoriamento Remoto [internet]; 2005 April 1621; Goiânia, Brazil. 2005 [access in 2017 January 13]. Available from: http://marte.sid.inpe.br/col/ltid.inpe.br/sbsr/2004/12.06.13.32/doc/473.pdf.

CAVACA, H. S. **Liminological Characterisation of Caraís Lagoon - ES.** Monograph, Department of Ecology and Natural Resources, Federal University of Espírito Santo. Vitória: 1992.

EASTMAN, J.R. **Idrisi Andes - Guide to GIS and Image Processing**. Clark Laboratory. Clark university. Worcester/MA. USA. 2006. 284p.

EMDEJA. **Preliminary Environmental Report of the Environmental Impact Studies of the Macrodrainage Project of the Lagoa Olho D'água** - Jaboatão River Estuary **Complex**. Jaboatão dos Guararapes, 2003.

EPIPHANIO, J. C. N.; GLERIANI, J. M.; FORMAGGIO, A. R.; RUDORFF, B. F. T. Vegetation indices in remote sensing of bean crops. **Pesquisa agropecuária brasileira**, Brasília, v. 31, n. 6, p. 445-454, 1996.

ESTEVES, F. A. **Coastal lagoons**: origin, functioning and management possibilities. In: Esteves, F.A. (ed). Ecology of the coastal lagoons of the Restinga de Jurubatiba National Park and the municipality of Macaé (RJ), 1998. p. 64.

FLORENZANO, T. G. Geotechnologies in applied geography: diffusion and access. **Revista do Departamento de Geografia**, 17 (2005), p. 24-29.

FRANÇA, A. F.; TAVARES JUNIOR, J. R.; MOREIRA FILHO, J. C.C. NDVI, NDWI and NDBI indices as tools for thematic mapping of the surroundings of the Olho **D'água** Lagoon, in Jaboatão dos Guararapes-PE. **IV Brazilian Symposium on Geodetic Sciences and Geoinformation Technologies**. UFPE Recife - PE, 06- 09 May 2012 p. 001 - 009.

FREITAS, R. **Entre mitos e limites**: as possibilidades do adensamento construtivo face à qualidade de vida no ambiente urbano. Recife: Ed. Universitária da UFPE, 2008.

GEHLEN VITÓRIA; **Beyond the Limits of the Urban:** Peri-Urban **Areas and the Environmental Question;** CADERNOS CERU, series 2, v. 21, n. 1, June 2010;

GILABERT, M.A; GONZÁLEZ-PIQUERAS; J; GARCÍA-HARO, F.J; MELIA, J. A generalised soil-adjusted vegetation index. **Remote Sensing of Environment.** v. 82. p. 303-310. 2002.

GOMES, S. C.; LIMA, E. S. de; ALHEIROS, M. M. Environmental diagnosis of the physical environment for the management of the Jaboatão River basin. **II Congress**

on the Planning and Management of Coastal Zones in Portuguese-Speaking Countries. IX Congress of the Brazilian Association of Quaternary Studies. II Quaternary Congress of Iberian-Speaking Countries. 2003.

GUERRA, A. T.; GUERRA, A. J. T. **Novo dicionário geológico- geomorfológico**. 9. ed. Rio de Janeiro: Bertrand Brasil, 648 p. 2011.

HERZOG P. CECILIA; **City for all:** (re) learning to live with nature; 1ª Ed; Rio de Janeiro; Mauad; 2013;

HUETE, A. R. Adjusting vegetation indices for soil influences. **International Agrophysics**, v.4, n.4, p.367-376, 1988.

JENSEN, J. R. (1949). **Remote sensing of the environment:** a perspective on terrestrial resources / translated by José Carlos Neves Epiphanio et al. São José dos Campos. SP. 2009.

KJERFVE, B. **Coastal Lagoon Processes**. Elsevier Oceanography Series, Amsterdam, 577p, 1984.
LEAL, J. P. **Geoenvironmental Study and Palaeographic Evolution of the Olho D'água Lagoon** (Jaboatão dos Guararapes). UFPE: 2002.
LIMA, Walter de Paula. **Forest hydrology applied to watershed management.** Piracicaba: University of São Carlos, ESALQ, Department of Social Sciences, 1996. 318 p. Mimeo.

LOPES, H.L.; ACCIOLY, L. J. O.; CANDEIAS, A. L. B.; SOBRAL, M. C. 2010. **Analysis of vegetation indices in the Brígida River Basin, Sertão do Estado de Pernambuco - III Brazilian Symposium on Geodetic Sciences and Geoinformation Technologies.** Recife - PE, 27-30 July 2010 p. 001 of 008.

LOUREIRO, D. D. **Evolution of heavy metals in Rodrigo de Freitas Lagoon.** RJ. 120f. Dissertation - Institute of Geosciences, Fluminense Federal University, Niterói, 2006.

MACEDO, O. G. de. **Evaluation of High Resolution Image Classifiers.** Urban area of Jaboatão dos Guararapes-PE. Dissertation (Master's Degree in Geodetic Sciences and Geoinformation Technologies) - Federal University of Pernambuco. 100p. 2010.

MARKHAM, B.L.; BARKER, L.L. Thematic mapper bandpass solar exoatmospherical irradiances. **International Journal of Remote Sensing**, v.8, n.3, p.517-523. 1987.

MARTINS, K. G.; **Disorganised Urban Expansion and Increased Environmental Risks to Human Health:** The Brazilian Case; Planaltina - DF, p. 10 - 17; 2012;

MELESSE, A. M., WENG, Q., THENKABAIL, P. S., SENAY, G. B. 2007. Remote Sensing Sensors and Applications in Environmental Resources Mapping and Modelling. **Sensors,** 7, 3209-3241.

MELLO, S. S. de. **Urban rivers and lakes:** urbanity and valorisation. 13th Meeting of the National Association for Postgraduate Studies and Research in Urban and Regional Planning. 25 to 29 May 2009. Florianópolis - Santa Catarina - Brazil.

NASCIMENTO, A. P. do. **Analysing the impacts of anthropic activities on coastal lagoons** - A case study of Lagoa Grande in Paracuru - CE. Dissertation. Federal University of Ceará. Fortaleza-CE, 2010.

OLIVEIRA, F. P. L. da.; BECKER, H. Limnological Characteristics of Lagoa do Sal - Coastal Plain of the Municipality of Beberibe - Ceará. **Revista de Geologia,** Vol. 19, no 2, 177 - 186, 2006.

OLIVEIRA, T. H.; SILVA, J. S.; MACHADO C. C. C.; GALVÍNCIO, J. D.; PIMENTEL, R. M. M.; SILVA, B. B. Moisture index (NDWI) and spatio-temporal analysis of surface albedo in the Moxotó river basin, PE. **Brazilian Journal of Physical Geography.** V.03, p.55 - 69, Homepage: www.ufpe.br/rbgfe. 2010.

PEREIRA, M. N., KURKDJIAN, M. L. N. O.; FORESTI, C. **Land cover and use through remote sensing.** São José dos Campos, Space Research Institute. 118 p. 1989.

PINHEIRO, C.; LARANJEIRA, M. Analysis of the thermal environment and ventilation conditions for the definition of climatic functions in the urban area of Guimarães. **Revista de Geografia e Ordenamento do Território** (GOT), no. 4 (December). Centre for Geography and Spatial Planning Studies, p. 249-272. 2013.

PONZONI, F. J. **Spectral Behaviour of Vegetation.** In: MENESES, P. R., NETTO, J. S. M. (org.) Remote sensing, target reflectance

natural. Brasília - DF: Editora Universidade de Brasília - UNB, Embrapa Cerrados, p 157-199, 2001.

PONZONI, F. J.; SHIMABUKURO, Y. E. **Remote Sensing in the Study of Vegetation.** São José dos Campos: Parêntese, 2010.

ROSENDO, J. dos S. **Vegetation indices and monitoring of land use and vegetation cover in the Araguari River Basin - MG - using data from the Modis sensor.** 2005. 130 p. Dissertation (Master's Degree in Geography and Territorial Management) - Postgraduate Programme in Geography, Federal University of Uberlândia, Uberlândia. 2005.

ROUSE, J.W.; HAAS, R.H.; SCHELL, J.A.; DEERING, D.W. **Monitoring**

vegetation systems in the great plains with ERTS. In: Third ERTS Symposium, Proceedings, NASA SP-351, NASA, Washington, DC, v. 1, p. 309-317, 1973.

SILVA, E. R. A. C.; MELO, J. G.; GALVINCIO, J. D. Identification of Areas Susceptible to Desertification Processes in the Middle Section of the Ipojuca Basin - PE through the Mapping of Vegetation Water Stress and the Estimation of the Aridity Index. **Revista Brasileira de Geografia Física,** Vol. 4, n 3, p. 629-649. 2011.

SILVA, E.R.A.C.; MIRANDA, R. Q.; FERREIRA, P. dos S.; GOMES, V. P. GALVÍNCIO, J.D. Estimation of Hydrological Stress in the Riacho do Pontal-PE Watershed. ISSN 2318-2962. DOI 10.5752/p.2318- 2962.2016v26n47p844. **Caderno de Geografia,** v.26, n.47, 2016.

TASUMI, M. **Progress in operational estimation of regional evapotranspiration using satellite imagery**. Idaho: University of Idaho, p. 378. 2003.

TUNDISI, J. G. **Água no Século XXI**: Enfrentando a escassez. São Paulo: Rima, 2009.

TUNDISI, J. G.; MATSUMURA-TUNDISI, T. **Limnologia**. Oficinas de Textos, 2008.

VIGANÓ, H. A; BORGES, E. F. FRANCA-ROCHA, W. J. S. **Analysing the performance of NDVI and SAVI vegetation indices from Aster images.** [Electronic version] Proceedings... XV Brazilian Symposium on Remote Sensing - SBSR, Curitiba, PR, Brazil, INPE p.1828. 2011.

VILLAÇA, F. **Espaço Intra-Urbano no Brasil**; 1ª Ed, São Paulo; FAPESP, 2001; WENG, Q.; QUATTROCHI, D. A. **Urban Remote Sensing.** CRC Press/Taylor and Francis, p. 448. 2006.

XAVIER, A. C.; SOARES, J. V. ALMEIDA, A.C. 2002. Variation of the Leaf Area Index in eucalyptus clones throughout their growth cycle. **Revista Árvore,** Viçosa-MG, v.26, n.4, p.421-427, 2002.
XU, H. **Extraction of Urban Built-up Land Features from Landsat Imagery Using a Thematic-oriented Index Combination Technique**. Photogrammetric Engineering & Remote Sensing Vol. 73, No. 12, pp. 13811391. December 2007.

ZHA, Y., GAO, J.; NI, S. Use of normalised difference built-up index in automatically mapping urban areas from TM imagery. **International Journal of Remote Sensing,** 24(3):583-594. 2003.

Printed by Books on Demand GmbH, Norderstedt / Germany